Math Concept Reader

Time to go Shopping

Developed for Harcourt, Inc., by Gareth Stevens, Inc. This edition published by Harcourt, Inc., by agreement with Gareth Stevens, Inc.

Printed in the United States of America

ISBN 13: 978-0-15-360286-3
ISBN 10: 0-15-360286-4

1 2 3 4 5 6 7 8 9 10 039 16 15 14 13 12 11 10 09 08 07

Time to go Shopping

by Jennifer Marrewa

Photographs by Kay McKinley

Chapter 1:
Who Wants to Write?

There is a visitor at school today. A writer talks to the class. He tells the children about the books he writes. He says that being an author is fun.

Von, Sara, and Jane love the stories the author shares. They want to write stories, too. They decide to start a writing club. They can make books together.

The next day they meet at Jane's house. They talk about the stories they will write. They talk about the pictures they will draw.

The friends need some supplies. Then they can start. They need pencils. They need an eraser. They need paper. Von thinks a big notebook is best.

Chapter 2:
The Writing Club Goes Shopping

Von makes a list of what they need. The children put money in their pockets. They are ready to go to the store. Jane's mom takes them.

The children look for the items they need. Von takes the list out of his pocket and reads it again. Then they walk through the store to find the items.

They find an eraser. It costs 56¢. Jane has two quarters. Sara has a nickel. Von has a penny. This makes 56¢. They will use these coins to buy the eraser.

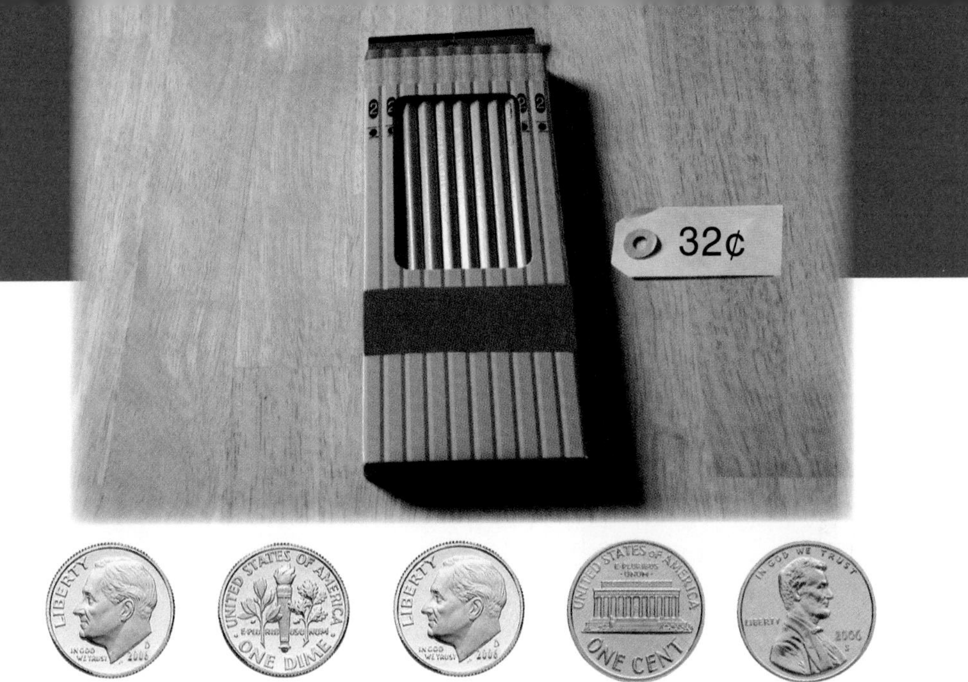

10¢, 20¢, 30¢, 31¢, 32¢

Sara finds the pencils. One box of pencils costs 32¢. Sara has three dimes. Von gives her a penny. This makes 31¢. Jane adds another penny. This makes 32¢. They will use these coins to buy the box of pencils.

Von finds the notebooks. He wants a blue one. It costs 86¢. Von has three quarters. Sara has one dime. Jane has one penny. This makes 86¢. They will use these coins to buy the blue notebook.

They walk to the counter and pay for their items. Now they have what they need for their writing club. Jane has a coin left over. Sara and Von have coins left, too.

Chapter 3: **More Coins to Count!**

The children look at their coins. Jane has one quarter left. Von has one quarter left, too. Sara has one quarter and one dime left.

25¢, 50¢, 75¢, 85¢

How much money do they have all together? They each put their coins on the counter. There are three quarters. There is one dime. They have 85¢ in all!

Chapter 4:
More Things to Buy

The children can buy something else for their writing club. What should they buy? They stop and think. They talk about it.

Von thinks they should buy colored pencils. A pack costs 95¢. The children look at their coins. There is not enough money to buy the colored pencils.

Sara thinks they should buy markers. The markers cost 90¢. The children look at their coins again. There is not enough money to buy the markers.

Jane thinks they should buy crayons. A box costs **85¢**. The children look at their coins one more time. There is enough money to buy the crayons.

The crayons will be perfect for drawing pictures. The children go to the counter. They pay for the crayons. Now they have everything they need.

They leave the store. They have pencils. They have a pink eraser. They have a blue notebook. They have crayons.

The children talk on the way home. They have lots of ideas for their stories. They have ideas for pictures, too.

Sara, Jane, and Von are ready to write. They write stories and draw pictures for their books. They are writers now! They are just like the visitor who came to school.

Glossary

eraser something used to rub out marks that are written or printed on paper

notebook sheets of paper bound together like a book

writer a person who writes books, stories, poems, and other things